The Pivotal Leader's Companion on the Untraveled Road

Self-Assessment Navigator

This is a Companion to the book "The Pivotal Tech Leader's Guide to the Untraveled Road"

1 | Page

Introduction

This companion to "The Pivotal Tech Leader's Guide to the Untraveled Road" is a self-assessment designed to help you evaluate your preparedness as a Pivotal Leader and navigate your path to, and on, the Untraveled Road.

Many who read the book told me that the content applied beyond technology leadership to include a broad spectrum of leadership across disciplines such as product management, business strategy, core capabilities (finance, etc.). Inspired by this I have pivoted myself to focus this assessment more broadly on the Pivotal Leader.

While this document does offer guidance it is not intended to be a development guide. Everyone's journey on the Untraveled Road is unique. "The Pivotal Tech Leader's Guide to the Untraveled Road" references many starting points and even experts in the many facets of becoming a Pivotal Leader. With the book as your guide, this assessment can help you navigate the journey.

The assessment is structured around the 3 Key Aspects of a Pivotal Leader:

- **A Prepared Mindset**: The mental approach that enables you to see possibilities others miss

- **Foundational Behaviors**: The actions that build trust, courage, and sustainable results

- **Leveled Up Skills**: The advanced capabilities that turn vision into reality

3 KEY ASPECTS TO A *PIVOTAL TECH LEADER*

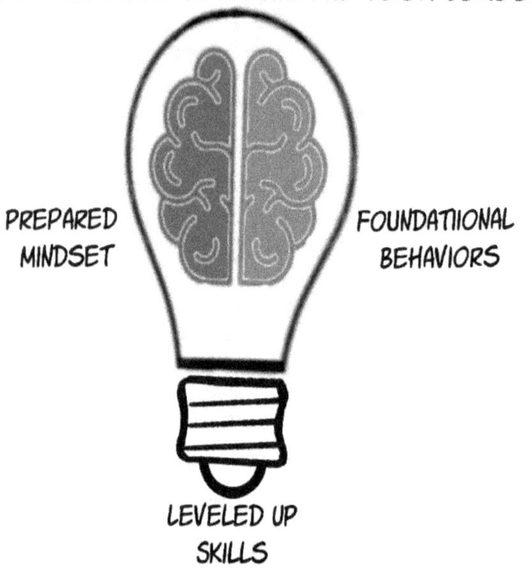

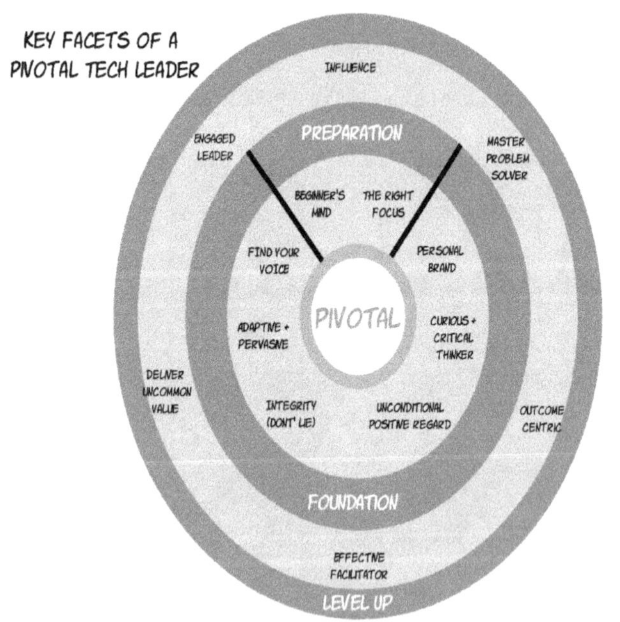

The Pivotal Leader's Companion on the Untraveled Road

The assessment engages you by prompting with questions, ratings and scenarios in each of the facets across all 3 Key Aspects.Pi

Each part represents one Key Aspect with chapters for each facet. Each chapter includes a 1-5 self-rating, diagnostic questions, scenarios, tables for assessment, and concrete action items.

How to Use This Assessment:

1. Answer honestly. This is for your development, not a performance review

2. Use the 1-5 scale where: 1=Rarely, 2=Sometimes, 3=Usually, 4=Often, 5=Always

3. Note specific examples for ratings of 1-2 (areas for growth) and 4-5 (strengths to leverage)

4. Revisit every 3-6 months to track your progress on the Untraveled Road

Part 1: A Prepared Mindset

The Principle: Pivotal Leaders must maintain clarity while navigating ambiguity, remain adaptable while staying focused, and inspire confidence while acknowledging uncertainty.

The Pivotal Leader's Companion on the Untraveled Road

General Mindset Questions

Rate yourself (1-5) on each:

☐ I maintain clarity of vision for my team/organization, especially when heading into unfamiliar territory

☐ I adapt quickly when faced with rapid change or unexpected challenges

☐ I proactively explore new technologies, industry trends, and potential disruptions

☐ I approach strategic planning with both decisiveness and openness to recalibration

☐ I respond to uncertainty and setbacks with emotional resilience

☐ I model resilience and optimism, inspiring others even during adversity

Reflection Questions:

- When was the last time you completely changed direction based on new information? What made you willing to pivot?

- What's your default response to hearing "that's never been done before"; excitement or concern?

- How do you balance confidence in your vision with openness to being wrong?

A Beginner's Mind

The Principle: Approaching familiar challenges with fresh eyes reveals solutions veterans miss. Cultivating curiosity over expertise creates breakthrough thinking.

Self-Assessment (Rate 1-5):

☐ I regularly question assumptions that "everyone knows" are true

☐ I seek perspectives from outside my industry or domain

☐ When facing a problem, I spend time asking "why" before jumping to "how"

☐ I'm comfortable saying "I don't know" or "teach me" in front of my team

☐ I actively look for ways my past success might be limiting my future thinking

☐ I create space for my team to challenge conventional wisdom without penalty

Critical Questions

- Do I dismiss new ideas because "that's not how we do things" or "I already know this"?

- Am I asking questions and exploring possibilities, or am I convinced there's only one "right" way?

- Can I adapt when new information challenges my existing beliefs, or do I resist change?

- Do I see myself as a lifelong student, or do I feel my knowledge makes me an expert preventing me from learning new things?

- Am I truly present and engaged in moments, or am I often just going through the motions, lost in thought?

Warning signs

☐ Feeling certain about things

☐ Struggling to learn new methods

☐ Becoming defensive when challenged

The Pivotal Leader's Companion on the Untraveled Road

Deep Dive Questions:

1. **When did your expertise last blind you to a better solution?**

   o   What was the situation?

   o   Who saw what you missed?

   o   What would "beginner's mind" have revealed sooner?

2. **Who on your team has the most unconventional background?**

   o   How often do you specifically seek their perspective?

   o   When was the last time their outsider view changed, or challenged, your thinking?

3. **What industry assumption would you eliminate if you could start over?**

   o   Why does that assumption persist?

   o   What's preventing you from challenging it now?

Action Items:

☐ Approach a familiar subject as if you've never encountered it, like a child learning.

☐ Ask "why" and "how" often, even about things you think you know well.

☐ Reflect on differences in opinions without judgment to find common ground.

☐ Identify one "industry standard" practice you'll question this month

☐ Schedule a learning session with someone outside technology

☐ Create a "beginner questions" practice for your next strategy meeting

The Right Focus

The Principle: Pivotal Leaders must master focus across four dimensions: themselves, outcomes, their team, and allies. Imbalance in any dimension creates blind spots.

Focus on Yourself

The Challenge: Self-awareness without self-absorption. Growth without narcissism.

Rate yourself (1-5):

- [ ] I have regular practices (journaling, coaching, reflection) for self-awareness

- [ ] I know my emotional triggers and manage them before they affect decisions

- [ ] I invest in my own learning and development as much as I expect from my team

- [ ] I can articulate my leadership philosophy and how it's evolved

- [ ] I recognize when I'm the bottleneck and actively work to remove myself

- [ ] I maintain physical and mental health practices that sustain performance

The Pivotal Leader's Companion on the Untraveled Road

Critical Questions:

1. **What's your leadership operating system?**

   - What principles guide your decisions under pressure?

   - When did you last update them based on failure or feedback?

2. **What percentage of your calendar is dedicated to your own growth?**

   - If it's less than 10%, what does that say about your priorities?

   - What would you need to say "no" to in order to invest in yourself?

3. **Who tells you the truth?**

   - Not just feedback; but also, the uncomfortable, ego-threatening truth?

   - How do you respond when they do?

Warning Signs (check any that apply):

☐ You're always the smartest person in the room

☐ You haven't significantly changed your mind about something important in 6+ months

☐ Your team waits for your opinion before sharing theirs

☐ You work more hours than your team does

☐ You can't remember the last time you admitted you were wrong

Focus on Outcomes

The Principle: Activity is not achievement. Pivotal Leaders measure what matters and manage to outcomes, not tasks. The Outcome-centric Metrics chapter in Part 3 assesses this focus further.

Rate yourself (1-5):

☐ I can clearly articulate the outcomes my team is accountable for

☐ My team knows the difference between their outputs and their outcomes

☐ I regularly challenge whether our current metrics actually predict success

☐ I prioritize ruthlessly, saying "no" to good opportunities to focus on great ones

☐ I track leading indicators, not just lagging results

☐ I celebrate outcomes achieved, not just effort expended

Assessment Questions:

1. **If your team had a perfect quarter by their current metrics but the company failed, what would that reveal about your outcome focus?**

2. **Can every person on your team connect their daily work to a measurable business outcome?**

   o   Test this: Ask three random team members today

   o   If they can't answer in 30 seconds, your outcomes aren't clear enough

3. **Are you measuring anything that doesn't actually matter?**

   o   What metrics exist because "we've always tracked them"?

   o   Which ones would you eliminate if you were starting fresh?

The Pivotal Leader's Companion on the Untraveled Road

Scenario Exercise:

Your team delivered everything on the roadmap on time. Customers are satisfied. Your metrics are green. But the CEO says your organization "isn't having enough impact."

- What does this reveal about your outcome focus?

- How would you respond?

- What would you change?

Action Items:

☐ List your team's top 5 metrics. For each one, explain exactly how it predicts business success

☐ Identify one "vanity metric" you'll stop tracking

☐ Create a "outcomes we're NOT pursuing" list and share it with your team and work with them to identify why not?

The Pivotal Leader's Companion on the Untraveled Road

Focus on Your Team

The Principle: Pivotal Leaders build teams that can navigate the Untraveled Road without them.

Rate yourself (1-5):

☐ I know each team member's career aspirations and actively support them

☐ I delegate authority, not just tasks; my team can make real decisions

☐ I create psychological safety where people take risks without fear

☐ I invest time in developing future leaders, even if they'll leave

☐ I celebrate team wins more than individual heroics

☐ My team would rather bring me problems early than wait until they're solved

Deep Assessment:

1. **If you left tomorrow, would your team thrive or struggle?**

   - Be honest: Are you building leaders or dependents?
   - What would fail without you? Why?

2. **When was the last time someone on your team made a significant decision without asking you?**

   - How did you respond?
   - Did they feel empowered or second-guessed?

3. **Who's the most talented person who's left your team?**

   - Why did they leave?
   - What does their departure tell you about your team focus?

The Pivotal Leader's Companion on the Untraveled Road

Team Health Diagnostic:

Answer Yes/No:

- ☐ Your team brings you issues before they're urgent

- ☐ Team members volunteer for difficult assignments

- ☐ Team members challenge your ideas in meetings

- ☐ You have successors identified for key roles

- ☐ Your best people are recruited by other parts of the company

- ☐ Failed experiments are discussed openly, not hidden

- ☐ People ask for stretch assignments, not just promotions

Score: 6-7 Yes = Strong team focus | 3-5 = Needs work | 0-2 = Critical gap

Red Flags (check any that apply):

- ☐ You're the only one who talks to senior leadership

- ☐ Your team asks permission for decisions they should own

- ☐ You know your team's technical skills better than their aspirations

- ☐ High performers leave for opportunities they could have had on your team

- ☐ You solve problems faster than developing problem-solvers

The Pivotal Leader's Companion on the Untraveled Road

Focus on Allies

The Principle: Pivotal Leaders understand that the hardest problems require coalitions, not heroics.

Rate yourself (1-5):

☐ I actively build relationships across organizational boundaries

☐ I invest time in understanding stakeholders' goals, not just my own needs

☐ I have strong relationships with peers in other functions (sales, marketing, operations)

☐ I create value for allies before asking for support

☐ I maintain relationships even when I don't need anything

☐ I leverage my network to create opportunities for my team

Leadership Network Assessment:

1. **Map your allies across three dimensions:**

    - Inside your organization (other departments, senior leaders, skip-levels)

    - Industry peers (competitors, partners, thought leaders)

    - Outside perspectives (advisors, board members, adjacent industries)

2. **For each category, answer:**

    - Who would take your call immediately?

    - Who would go out of their way to help you?

    - Who have you helped recently without expecting return?

3. **When was the last time you:**

    - Made an introduction that benefited someone else?

    - Asked for advice (not approval) from a peer?

    - Collaborated with another team to solve their problem?

The Pivotal Leader's Companion on the Untraveled Road

Ally Strategy Exercise:

Think of your most ambitious initiative for next quarter.

1. Who needs to succeed for your initiative to work?

2. What's in it for them if you succeed?

3. What obstacles are they facing that you could help with?

4. How strong is your relationship with each key ally?

Warning Signs:

- ☐ You only reach out to people when you need something

- ☐ Your peer relationships are transactional, not genuine

- ☐ You compete with other leaders rather than collaborating

- ☐ Your network consists mostly of people like you

- ☐ You can't recall the last time you introduced two people who could help each other

Action Items:

- ☐ Identify your three most important allies and schedule time with each

- ☐ Make one introduction this week that benefits others, not you

- ☐ Find one way to support a peer's initiative before asking for their support

Part 2: Foundational Behaviors

The Principle: Behaviors are how your mindset becomes visible. Pivotal Leaders must consistently demonstrate the behaviors that build trust, enable truth, and drive results, especially when it's uncomfortable.

General Behavioral Questions

Rate yourself (1-5):

☐ I consistently uphold ethical standards and communicate the importance of integrity

☐ I foster a culture of trust and inclusion, welcoming diverse perspectives and feedback

☐ I effectively empower my team, delegate authority, and encourage autonomy

☐ I create a safe environment for experimentation, risk-taking, and learning from failure

☐ I nurture talent, provide development opportunities, and support growth

☐ I'm transparent about challenges and failures, and hold myself and others accountable

Reflection:

- Which behavior is your strongest foundation? How do you know?

- Which behavior needs the most work? What's the cost of not improving it?

- What behavior would your team say you demonstrate least consistently?

Craft Your Brand

The Principle: Your brand is what makes you memorable and distinctive. Pivotal Leaders intentionally craft their brand around their "superpower," communicate it clearly, and evolve it as they grow.

Rate yourself (1-5):

☐ I can clearly articulate what makes me uniquely valuable as a leader

☐ I have a compelling "7-second pitch" that captures my leadership brand

☐ My brand is based on patterns of success, not just tactical achievements

☐ I actively expose and test my brand through networking and engagement

☐ I regularly evolve my brand to stay relevant and aligned with my growth

☐ I'm conscious of how my brand is perceived and seek feedback to refine it

The Pivotal Leader's Companion on the Untraveled Road

Brand Foundation Assessment:

Your Superpower:

1. **What patterns have led to your success?**

   o   List 3-5 situations where you've consistently excelled

   o   What's the common thread?

   o   What capability do people consistently seek you out for?

2. **Complete this sentence:** "I am uniquely great at

   _______________________________________" If you can't complete it in one clear sentence, your brand isn't focused enough.

3. **Your tactical resume vs. your brand:**

   o   Tactical: "I know Python, led a team of 20, shipped 15 features"

   o   Brand: "I turn chaotic technical debt into strategic advantage"

Which sounds more like you talk about yourself?

The 7-Second Pitch Test:

Write your current 7-second pitch:

____

Now test it:

- Does it communicate what you what you achieve?

- Is it memorable (would someone repeat it)?

- Does it differentiate you from other tech leaders?

- Does it make people want to learn more?

If you answered "no" to any of these, your pitch needs work.

Brand Exposure Assessment:

In the past 3 months:

- ☐ Have you attended networking opportunities (online or in-person)?

- ☐ Have you engaged in conversations outside your immediate team?

- ☐ Have you provided advice or answered questions for others?

- ☐ Has anyone reached out to you based on your reputation?

- ☐ Have you been memorable enough that someone contacted you later?

Score: 4-5 Yes = Strong brand exposure | 2-3 = Needs work | 0-1 = Hidden brand

Brand Stereotype Risk:

What are you known for that might be limiting?

Common tech leader stereotypes:

- ☐ The firefighter (always in crisis mode)

- ☐ The technologist (great with tech, not with people)

- ☐ The executor (ships on time, but not strategic)

- ☐ The innovator (lots of ideas, questionable follow-through)

If you checked one:

- Is this stereotype accurate?

- Is it limiting your opportunities?

- How can you leverage it while evolving beyond it?

Example: Being known as a crisis manager gets attention, but don't let it become your only brand. Add strategic planning or team development to round it out.

Brand Evolution Check:

Answer honestly:

1. **When did you last update how you describe yourself?**

    o   Last month: Evolving appropriately

    o   Last year: May be getting stale

    o   Can't remember: Brand is stagnant

2. **Does your brand reflect who you are today or what you did 3 years ago?**

    o   If it's 3 years ago, you're trading on past reputation

3. **Does your brand align with where you want to go?**

    o   If not, what needs to change?

Brand Fit Assessment:

Sometimes you've crafted the right brand but you're in the wrong environment.

Ask yourself:

- Am I in the right place to make a difference with this brand?

- Does my organization value what I'm uniquely great at?

- Am I in a culture that fits my brand?

If you answered "no" to these, you might need to find a better environment, not change your brand.

Scenarios:

Scenario 1: You're known as "the person who fixes technical debt" but you want to be seen as a strategic innovator.

- How do you leverage your current brand?

- What do you need to add to evolve it?

- What opportunities would demonstrate the evolution?

Scenario 2: People keep bringing you crisis situations because you're good at firefighting, but it's exhausting.

- Is crisis management part of your brand?

- How do you maintain the positive (reliable, capable) while setting boundaries?

- What would you rather be known for?

Scenario 3: You realize your brand is "tactical executor" but leadership wants strategic thinkers.

- Can you reframe your tactical excellence as strategic execution?

- What strategic behaviors do you need to add?

- How do you signal the change?

Brand Refinement Strategy:

Based on feedback and self-reflection:

1. **What should I continue doing?** (Reinforces positive brand)

2. **What should I stop doing?** (Undermines desired brand)

3. **What should I start doing?** (Builds toward future brand)

Action Items:

- Write and test your 7-second pitch with 3 people this week

- Identify one networking opportunity this month to expose your brand

- Get brand feedback: Ask 3 colleagues "What am I known for?"

- Assess environment fit: Am I in the right place for my brand to thrive?

Own the Conversation

The Principle: Pivotal Leaders don't wait for others to frame discussion; they own the conversation by establishing the language, aligning stakeholders, and connecting capabilities to business outcomes.

Rate yourself (1-5):

☐ I frame technology, or other domain knowledge, conversations in business outcome terms, not jargon

☐ I establish shared language that enables cross-organizational alignment

☐ I proactively communicate value before stakeholders ask

☐ I connect technical achievements to business impact consistently

☐ I set the agenda for domain conversations rather than reacting to others' framing

☐ I'm comfortable leading conversations with non-domain centric executives

Communication Mastery Assessment:

The "Moving the Needle" Test:

Think of your most recent significant initiative.

Can you articulate:

- **The business outcome** (not the domain's output)?

- **Why it matters** to executives who don't understand the domain jargon?

- **The value in their language** (revenue, cost, risk, capability)?

If you defaulted to domain details instead of business outcomes, you're not owning the conversation.

Conversation Ownership Diagnostic:

Who typically frames technology discussions in your organization?

- You set the agenda and language

- Business stakeholders frame it, you respond

- It varies by situation

- Product managers or others frame it

If you're not consistently the one framing d domain conversations, you're not owning them.

Language Alignment Check:

Do you have frameworks or tools that enable consistent language across:

Stakeholder Group Shared Language Exists? What Framework/Tool?

Business Leadership Yes / No

Operations   Yes / No

Technology Teams   Yes / No

External Partners   Yes / No

If you answered "No" to any: You're missing critical alignment infrastructure.

Examples of language alignment tools:

- TOGAF/ArchiMate for enterprise architecture

- Business capability maps

- Value stream mapping

- OKR frameworks that connect tech to business

Finding Your Voice:

Rate your capability (1-5) on each communication essential:

☐ **Brand credibility**: I leverage my brand to establish authority in conversations

☐ **Outcomes first**: I lead with business outcomes before discussing technical tactics

☐ **Planning ahead**: I prepare for important conversations with clear topics and priorities

☐ **Clear and concise**: I communicate without unnecessary technical jargon

☐ **Know your audience**: I adapt my message based on who I'm talking to

☐ **Adaptability**: I read the room and adjust my approach mid-conversation

☐ **Comprehensive feedback**: I use active listening and probing questions to ensure understanding

☐ **Keep it moving**: I set boundaries and maintain focus on productive outcomes

The Office Space Problem:

Are you perpetuating the "engineers aren't good with people" stereotype?

Self-assessment:

- Do business stakeholders bypass you to talk to product managers about technology?

- Are you more comfortable in technical forums than business meetings?

- Do you default to technical explanations when asked business questions?

- Have you been told you need to "speak business language"?

If you checked any boxes: You're not owning the conversation, someone else is translating for you.

Conversation Ownership Scenarios:

Scenario 1: An executive asks "Are we getting value from our cloud migration?"

- **Poor response**: "We've migrated 60% of workloads and reduced latency by 200ms"

- **Owning it**: "With our shared understanding of value, here are the areas we are, and are not, attaining value: _________________________________. Are there other value propositions you have in mind?"

Your typical response? ____

Scenario 2: A business leader says "We need to move faster with AI"

- **Poor response**: "We're evaluating LLM frameworks and setting up ML pipelines"

- **Owning it**: "Let's talk about which business outcomes you're trying to accelerate. Once we align on the target, I can show you 3 paths with different speed/risk tradeoffs."

Your typical response? ____

Scenario 3: In a strategy meeting, someone suggests a technology approach you know won't work or for which there are substantial unknowns.

- **Poor response**: Stay silent or say "that won't work technically"

- **Owning it**: "Let me make sure I understand the outcome you're aiming for:

  ____

  Now, here's are the unknowns and risks we need to address:

  ____

  We should make sure we have a clear understanding of the problem definition, desired outcomes and the value proposition before pursuing solutions."

Your typical response? ____

The Pivotal Leader's Companion on the Untraveled Road

Enterprise Architecture & Language Ownership:

If you're in an EA role or have EA responsibilities:

☐   Do you have a consistent framework for aligning WHO, WHAT, WHERE, WHEN, WHY, and HOW across business, operations, and technology?

☐   Can you model initiatives from strategy through delivery without repeatedly bootstrapping alignment conversations?

☐   Do business stakeholders see you as essential to strategic technology conversations?

☐   Have you reduced the number of alignment meetings needed for major initiatives?

EA-specific warning signs:

- Your architecture artifacts are only understood by other architects
- Business leaders see architecture as a compliance checkpoint, not a value enabler
- You're great at technology architecture but weak at business architecture
- Your frameworks are bespoke rather than based on standards that enable collaboration

Owning the Commercialization Conversation:

When your organization evaluates emerging technologies:

☐   Do you own the conversation about applying appropriate perspective lenses?

☐   Do you cut through hype confetti and technical jargon to focus on value bridges?

☐   Do you coordinate between technology curators, investors, solution providers, and internal stakeholders?

☐   Do you establish the language that connects technology capability to business outcomes?

If you're not doing these things, someone else is owning the conversation, or no one is, which is worse.

The Pivotal Leader's Companion on the Untraveled Road

Communication as Core Skill:

For yourself:

- I invest in improving my communication skills as much as my technical skills

- I seek feedback on my communication effectiveness

- I practice important conversations before they happen

- I learn from communicators I admire

For your team:

- I develop communication skills in my team members

- I create opportunities for my team to practice stakeholder communication

- I give feedback on how they communicate, not just what they deliver

- I model effective communication in my own behavior

Action Items:

- Reframe your next technical update in pure business outcome terms—no jargon

- Identify one technology conversation where you need to establish the framing

- If you lack language alignment tools, research frameworks like TOGAF/ArchiMate or value stream mapping

- Practice your response to "What value is technology providing?" in 60 seconds

Curiosity and Critical Thinking

The Principle: Curiosity and critical thinking are the yin and yang of collaboration. Pivotal Leaders foster constructive tension where all perspectives are heard, curiosity drives exploration, and critical thinking ensures rigor, without defensiveness destroying the dialogue.

Rate yourself (1-5):

☐ I'm genuinely curious about criticism and feedback, not defensive

☐ I create environments where constructive tension produces better outcomes

☐ I balance curiosity (exploring possibilities) with critical thinking (evaluating rigor)

☐ I solicit criticism positively: "What could I be doing better?"

☐ I ensure criticism is tied to evidence, not just opinion or judgment

☐ I foster honest feedback loops where criticism leads to action, not just venting

Curiosity About Criticism:

Your default response to criticism:

When someone criticizes your decision or approach, which happens first?

- I get defensive and explain why they're wrong

- I feel hurt or attacked

- I get curious about what I'm missing

- I ask questions to understand their perspective

- I look for the kernel of truth in their feedback

If you checked the first two, defensiveness is blocking your growth.

The Constructive Tension Assessment:

Rate your team's capability (1-5):

☐ **Diverse perspectives engaged**: All viewpoints are explored, not just consensus

☐ **Productive disagreement**: People challenge ideas without personal attacks

☐ **Evidence-based feedback**: Criticism is rooted in facts, not opinions

☐ **Action-oriented**: Feedback leads to decisions and changes, not just discussion

☐ **Psychological safety**: People share concerns without fear of retaliation

☐ **Managed stress**: Tension is productive, not overwhelming

Constructive Tension vs. Toxic Conflict:

Constructive Tension (Good):

- Focuses on ideas and outcomes
- Evidence-based disagreement
- Leads to better decisions
- Everyone feels heard
- Results in action

Toxic Conflict (Bad):

- Personal attacks
- Opinion-based judgment
- Leads to paralysis or forced consensus
- Some voices silenced
- Breeds resentment

Which describes your team's typical disagreements?

The "Happy Shiny People" Problem:

- ☐ Does your team agree too easily?
- ☐ Do people rarely challenge your ideas?
- ☐ Is consensus reached quickly without real debate?
- ☐ Do people seem too comfortable with the status quo?

If you checked these boxes: You don't have enough constructive tension. Your team isn't challenging each other enough to produce breakthrough thinking.

The Pivotal Leader's Companion on the Untraveled Road

Stress Level Assessment:

Constructive tension requires optimal stress, not too little, not too much.

Signs of insufficient stress (too comfortable):

- People accepting status quo without questioning
- No urgency around problems
- Low energy in discussions
- Avoiding difficult decisions

Signs of excessive stress (overwhelmed):

- Behavioral changes (withdrawal, aggression, burnout)
- Paralysis instead of action
- People playing it safe to avoid failure
- High turnover or absenteeism

Your team's current state: Too comfortable / Optimal / Overwhelmed

Curiosity & Critical Thinking Balance:

For each situation, which do you need more of:

Situation	Need More Curiosity	Need More Critical Thinking
Exploring new technology		
Evaluating technical debt		
Designing architecture		
Responding to customer complaints		
Assessing vendor proposals		
Planning team growth		

Pattern check:

- Do you over-index on one consistently?

- Where's your blind spot?

Leader's Achilles Heel: Aversion to Honesty:

Self-assessment:

☐ Do people tell me what I want to hear more than what I need to hear?

☐ Does feedback about me feel confrontational rather than helpful?

☐ Do I react defensively when my decisions are questioned?

☐ Is criticism of me seen as risky by my team?

☐ Do I address criticism by explaining/justifying rather than listening/acting?

If you checked any boxes: You have an honesty aversion problem that's limiting your team's effectiveness.

The root cause: Feedback is delivered confrontationally OR received defensively (or both).

Honest Critique vs. Confrontation:

Confrontational (Bad):

- Based on opinion and judgment

- Attacks person, not issue

- No path to improvement offered

- Focused on blame

Honest Critique (Good):

- Based on facts and evidence

- Focuses on behavior/outcome

- Includes suggestions for improvement

- Focused on growth

How is feedback typically delivered on your team?

Creating Evidence-Based Feedback:

Poor feedback: "Your communication skills aren't good enough"

Evidence-based feedback: "In the last three stakeholder meetings, we lost the audience when you went deep on technical details instead of leading with business outcomes. Can we work on framing?"

Practice: Take a piece of feedback you need to give and make it evidence-based:

Opinion-based version: ____

Evidence-based version: ____

Fostering Curious & Critical Teams:

Do you model:

- Asking "What am I missing?" instead of defending your position

- Soliciting criticism: "What could I be doing better?"

- Changing your mind based on evidence

- Acknowledging when you were wrong

- Thanking people for challenging your thinking

Do you create:

- Forums for debate and challenge

- Norms that separate idea from person

- Processes that require evidence for claims

- Action plans based on feedback

- Recognition for productive disagreement

Scenarios:

Scenario 1: A team member criticizes your technical strategy in a public meeting.

- **Defensive response**: Explain why they don't understand, defend your decision

- **Curious response**: "Tell me more about your concern. What am I not seeing?"

How would you actually respond? ____

Scenario 2: Your team always agrees with you quickly, no pushback.

- What does this signal?

- How do you create more constructive tension?

- What question could you ask to invite challenge?

Scenario 3: Two senior engineers are in heated disagreement. It's getting personal.

- How do you intervene?

- How do you redirect to constructive tension?

- What evidence do you ask for?

Beginner's Mind + Critical Thinking:

The combination is powerful:

- **Beginner's Mind** asks: "What if we're wrong about our assumptions?"

- **Critical Thinking** asks: "What evidence supports or refutes our hypothesis?"

Together they create: Curious exploration grounded in rigorous evaluation.

Do you practice both?

Action Items:

- Next time you receive criticism, respond with "Tell me more" instead of explaining

- Assess your team's stress level; adjust to create optimal constructive tension

- Establish one norm for evidence-based feedback in your team

- Identify one area where you need more curiosity, one where you need more critical thinking

- Ask your team: "What am I doing that makes it hard to give me honest feedback?"

Professional Courage

The Principle: Courage is choosing the right action over the comfortable one. Pivotal Leaders must consistently demonstrate courage in decisions, conversations, and commitments.

The Principle:
Courage is choosing the right action
over the comfortable one.

Rate yourself (1-5):

☐ I make difficult decisions without waiting for consensus or perfect information

☐ I advocate for what's right even when it's politically risky

☐ I have difficult conversations promptly rather than avoiding them

☐ I speak up when I disagree with leadership, even when it's uncomfortable

☐ I take responsibility for failures publicly, not just privately

☐ I defend my team when they take smart risks that don't work out

The Pivotal Leader's Companion on the Untraveled Road

Courage Audit:

1. **What's the most courageous thing you've done as a leader in the past 6 months?**

   - What was at risk?

   - What made it hard?

   - What happened?

2. **What difficult conversation are you currently avoiding?**

   - Why are you avoiding it?

   - What's the cost of continued avoidance?

   - What would courage require you to do this week?

3. **When was the last time you changed your position based on someone challenging you?**

   - How did you respond in the moment?

   - How did you communicate the change?

Scenario Tests:

How would you respond?

Scenario 1: Your team delivers a project on time, but you realize the original strategy was flawed. Leadership is celebrating the execution.

- Do you publicly acknowledge the strategic error?

- How do you balance team morale with honesty about the outcome?

Scenario 2: A peer leader is making a decision you believe will harm the company. They outrank you and have the CEO's support.

- Do you voice your concern?

- How do you raise it without appearing territorial or undermining?

Scenario 3: Your best engineer made a costly mistake due to a reasonable calculated risk.

- How do you respond publicly?

- How do you respond privately?

- What do you say to leadership?

Courage Development:

☐ Identify one avoided conversation you'll have this week

☐ Find one situation where you'll publicly take responsibility

☐ Practice one instance of disagreeing with someone senior

Integrity (Don't Lie)

The Principle: Integrity isn't just honesty; it's the alignment between what you say, what you believe, and what you do. Pivotal Leaders keep promises, tell hard truths, and are the same person in every room.

The Principle: Integrity isn't just honesty; it's the alignment between what you say, what you believe, and what you do.

Rate yourself (1-5):

☐ I tell the truth even when it's inconvenient or makes me look bad

☐ My commitments match my capacity; I don't over-promise

☐ I keep promises, or renegotiate them proactively when circumstances change

☐ I'm the same leader in front of executives as I am with my team

☐ I admit mistakes quickly and completely

☐ I say "I don't know" rather than speculating or faking expertise

Integrity Check:

1. **What have you said you would do that you haven't done?**

   - Make a list of unkept commitments from the past 3 months

   - For each: Did you renegotiate, or just let it drop?

   - What does this pattern tell you?

2. **When was the last time you:**

   - Said something in an executive meeting you wouldn't say to your team?

   - Let someone believe something you knew wasn't quite true?

   - Agreed to a deadline you knew was unrealistic?

3. **What would your team say about your integrity?**

   - Would they say you keep your word?

   - Do they believe you when you explain decisions?

   - Do they trust you to tell them hard truths?

The Integrity Stress Test:

Read each scenario and write your response:

Scenario 1: You're asked to commit to a timeline you believe is impossible. The executive asking wants to hear "yes" to support their board presentation.

Scenario 2: A team member asks if layoffs are coming. You've been told confidentially that they are, but not to communicate yet.

Scenario 3: You discover your predecessor misrepresented the state of the technology you now own. Leadership is making decisions based on the misrepresentation.

Integrity versus Loyalty:

When these conflict, which wins?

- Loyalty to your team vs. truth to leadership?

- Loyalty to your boss vs. what's right for the company?

- Your reputation vs. your team member's mistake?

Action Items:

- ☐ Review all commitments made in the past month; identify any you need to renegotiate

- ☐ Identify one truth you've been softening; commit to stating it clearly

- ☐ Ask a trusted peer: "Is there anywhere my words and actions don't match?"

Unconditional Positive Regard

The Principle: Separating person from performance. Believing in people's potential even when critiquing their current results. Pivotal Leaders create psychological safety through genuine respect for every individual.

Rate yourself (1-5):

☐ I critique work and decisions without attacking character or intelligence

The Pivotal Leader's Companion on the Untraveled Road

☐ I assume positive intent until proven otherwise

☐ I invest in developing people even when their current performance disappoints

☐ I create space for people to fail without losing my confidence in them

☐ I see potential in people others have written off

☐ My feedback focuses on growth, not judgment

Assessment Questions:

1. **Who on your team do you find most difficult?**

    o   Can you articulate what you genuinely respect about them?

    o   When did you last tell them something you appreciate?

    o   Do they know you believe in their potential?

2. **When someone on your team fails, what's your first thought?**

    o   "What's wrong with them?" (person focus)

    o   "What went wrong?" (situation focus)

    o   "What did I miss in setting them up for success?" (system focus)

3. **How do you talk about team members who aren't in the room?**

    o   Would you say the same things with them present?

    o   Do you defend them or join in criticism?

    o   Do you share their wins as enthusiastically as their failures?

The UPR Challenge:

Think of someone on your team whose performance is struggling.

Complete these sentences:

- "What I genuinely appreciate about [name] is..."

- "The potential I see in them is..."

- "The ways I may have contributed to their struggle include..."

- "What I'm committed to doing differently to support them is..."

Scenario Practice:

Scenario 1: An engineer ships a feature with a critical bug that takes the site down for 2 hours.

- Your instinctive response?

- What do you say in the post-mortem?

- How do you ensure they still feel trusted to take risks?

Scenario 2: A direct report is consistently missing deadlines. They're technically brilliant but seem unmotivated.

- How do you open the conversation?

- How do you separate care for them from consequences for performance?

Scenario 3: Someone applies for a promotion you don't think they're ready for.

- How do you deliver that message with respect?

- How do you keep them engaged rather than demoralized?

Red Flags (check any that apply):

☐ You mentally write people off after repeated disappointments

☐ You have "problem children" you complain about to other leaders

☐ Your tone changes noticeably between high and low performers

☐ You avoid giving developmental feedback because "they can't handle it"

☐ You've stopped investing in someone's growth because "they'll never get it"

Action Items:

☐ Identify someone you've mentally written off; find one thing to genuinely appreciate and tell them

☐ Review your last 10 pieces of feedback; ratio of growth-focused vs. judgment-focused?

☐ Practice reframing: Take one person-focused criticism and restate it as a situation-focused observation

Adaptive & Persuasive Leadership

The Principle: One leadership style doesn't fit all situations. Pivotal Leaders diagnose what each moment requires and adjust their approach, while maintaining their core values.

Rate yourself (1-5):

☐ I adjust my leadership style based on the person, situation, and context

☐ I can be directive when needed and collaborative when appropriate

☐ I read the room accurately and adapt my communication accordingly

☐ I influence through multiple approaches; data, story, vision, relationship

  Self-Assessment Navigator

The Pivotal Leader's Companion on the Untraveled Road

☐ I know when to push and when to listen

☐ I maintain authenticity while being flexible in approach

Leadership Flexibility Assessment:

When do you default to:

Directive (telling): _____% of the time

Coaching (asking): _____% of the time

Delegating (empowering): _____% of the time

Consensus (facilitating): _____% of the time

Should total 100%

Now answer:

1. Is this distribution intentional or habitual?

2. What situations call for each style?

3. Which style do you over-use? Under-use?

Persuasion Toolkit:

Rate your effectiveness (1-5) with each persuasion approach:

☐ **Data & Logic**: Building the rational case with evidence

☐ **Vision & Inspiration**: Painting the future state

☐ **Relationship & Trust**: Leveraging credibility and connection

☐ **Story & Narrative**: Making it memorable and relatable

☐ **Political & Strategic**: Understanding power dynamics and timing

☐ **Urgency & Consequences**: Highlighting what's at stake

Which is your go-to? Which do you avoid?

The Pivotal Leader's Companion on the Untraveled Road

Adaptive Leadership Scenarios:

For each scenario, identify the best leadership style and why:

Scenario 1: Critical production issue, customers affected, team stressed

- Best style?

- Why?

- What would you say?

Scenario 2: Team debate about technical architecture, smart people disagree

- Best style?

- Why?

- How do you facilitate?

Scenario 3: Junior engineer with potential but needs development

- Best style?

- Why?

- What's your approach?

Scenario 4: Executive asking for timeline you think is aggressive but possible

- Best style?

- Why?

- How do you respond?

Persuasion Practice:

Think of something you need to convince leadership to support.

Now craft three different versions of your pitch:

1. **Data-driven** version (for analytical leaders)

2. **Vision-driven** version (for strategic leaders)

3. **Risk-focused** version (for cautious leaders)

Which one is most natural for you? Which is hardest? Who on your stakeholder list needs which version?

Action Items:

☐ Identify one situation this week where you'll consciously use a non-default leadership style

☐ Map your key stakeholders by what persuades them—adjust your approach accordingly

☐ Get feedback: Ask three people how they experience your leadership style

---

Part 3: Leveled Up Skills

The Principle: Mindset and behaviors create the foundation, but Pivotal Leaders need advanced skills to execute in uncharted territory. These aren't generic leadership skills; they're the capabilities that separate good leaders from those who create entirely new paths (The Untraveled Road) for themselves and their teams.

General Skills Questions

Rate yourself (1-5):

☐ I've acquired advanced skills specifically to address current and future challenges

☐ I leverage these skills to champion innovation and implement new solutions

The Pivotal Leader's Companion on the Untraveled Road

☐ I allocate resources to exploratory projects and evaluate success with clear criteria

☐ I'm committed to lifelong learning through mentorship, coaching, or formal education

☐ I incorporate feedback to refine my approach and model a growth mindset

☐ I regularly share lessons from my experiences to mentor others

Reflection:

- What skill have you developed in the past year that directly enabled new capability?

- What skill do you need next that you don't currently have?

- How are you systematically developing your team's skills, not just your own?

Influence

The Principle: Pivotal Leaders create change without formal authority. Influence is the ability to shape decisions, align stakeholders, and move organizations, regardless of your position.

Rate yourself (1-5):

☐ I can influence peers who don't report to me

☐ I shape decisions before they're made, not just react after

☐ I build coalitions across organizational boundaries

☐ I influence up (changing leadership's perspective)

☐ I influence laterally (gaining peer support)

☐ I influence down (enrolling teams in vision)

Influence Inventory:

Map your influence across organizational levels:

Level	Strength (1-5)	Evidence	Gap to Close
Executive/C-Suite			
Peer Leaders			
Skip-Level Teams			
Cross-functional Partners			
External Stakeholders			

Influence Diagnostic:

1. **When was the last time you changed someone's mind about something important?**

   - Who was it?

   - What approach did you use?

   - What made it work?

2. **Think of a decision that went against your recommendation:**

   - Why didn't you successfully influence the outcome?

   - Was it your argument, your timing, your stakeholders, or your credibility?

   - What would you do differently?

3. **Who influences you? Why?**

   - What makes their influence effective?

   - What can you learn from their approach?

Influence Scenarios:

Scenario 1: You need budget for an exploratory project with uncertain ROI. The CFO is skeptical.

- How do you build your case?

- Who else needs to be enrolled first?

- What's your influence strategy?

Scenario 2: A peer leader's decision will create problems for your team. They outrank you.

- How do you raise the issue without seeming territorial?

- What allies do you need?

- What's your opening conversation?

Scenario 3: Your team is resistant to a change you believe is necessary.

- How do you enroll them rather than command them?

- What resistance do you need to understand vs. overcome?

- How do you build momentum?

Action Items:

- ☐ Identify one decision you want to influence; map all stakeholders and what matters to each

- ☐ Practice pre-wiring: Talk to key stakeholders before the decision meeting

- ☐ Build one new relationship with someone who can expand your influence

---

Problem Solving

The Principle: Pivotal Leaders don't just solve problem; they solve the right problems in ways that create lasting value and build capability.

Rate yourself (1-5):

☐ I invest time in problem definition before jumping to solutions

☐ I distinguish between symptoms and root causes

☐ I involve the right people in problem-solving (those closest to the problem)

☐ I create frameworks that help others solve similar problems

☐ I know when to solve vs. when to facilitate others solving

☐ I balance speed with thoroughness based on problem criticality

Problem-Solving Assessment:

Your Problem-Solving Profile:

When facing a problem, I typically (rank 1-4):

☐ Jump to solution mode immediately

☐ Analyze deeply before acting

☐ Gather the team to brainstorm

☐ Look for analogous problems solved before

None are wrong, but:

- Do you over-use one approach?

- Which situations call for which approach?

- What's your blind spot?

Problem Definition Test:

Think of a current challenge. Complete these sentences:

Surface problem: "We need to..."

Underlying problem: "The real issue is..."

Root cause: "This happens because..."

Systemic cause: "The system that creates this is..."

If you skipped straight to solution, you're solving the wrong problem.

Problem-Solving Scenarios:

Scenario 1: Deadline slipping, team working long hours, morale suffering

- What's the problem you'd solve?
- How do you know it's the right problem?
- Who should be involved in solving it?

Scenario 2: Customer complaints increasing, but specific issues vary widely

- What's your problem-solving approach?
- How do you avoid solving one symptom at a time?
- How do you involve the team?

Scenario 3: Two senior engineers propose completely different technical approaches

- Is this a problem to solve or a decision to make?
- How do you approach it?
- What's your role?

Problem-Solving Capability Building:

☐ Do you solve problems or teach problem-solving?

☐ Can your team solve problems without you?

☐ Have you created frameworks that outlive individual problems?

☐ Do you celebrate problem discovery as much as problem-solving?

Action Items:

- [ ] For your next problem, spend 2x normal time on definition before solution

- [ ] Identify one recurring problem; create a framework others can use

- [ ] Next time someone brings you a problem, ask "what have you tried?" before solving it

Facilitation

The Principle: Pivotal Leaders get better outcomes by unlocking the collective intelligence of the room rather than being the smartest person in it.

Rate yourself (1-5):

☐ I design meetings with clear outcomes and appropriate structures

☐ I create space for quiet voices while managing dominant ones

☐ I know when to facilitate vs. when to participate

☐ I can run effective workshops for strategy, problem-solving, or decision-making

☐ I handle conflict productively without avoiding or escalating

☐ I capture outcomes and ensure follow-through

Facilitation Skills Inventory:

Rate your capability (1-5) with each:

☐ **Meeting Design**: Clear purpose, right participants, appropriate length

☐ **Process Selection**: Choosing techniques (brainstorm, debate, consensus, etc.)

☐ **Participation Management**: Drawing out introverts, managing extroverts

☐ **Conflict Navigation**: Productive disagreement without toxicity

☐ **Decision-Making**: Moving from discussion to decision effectively

☐ **Energy Management**: Reading and shifting room dynamics

☐ **Capture & Follow-up**: Documenting and ensuring action

Facilitation Audit:

Review your last 5 meetings you led:

1. **Did you talk more than anyone else?**

   o If yes, were you facilitating or dominating?

  Self-Assessment Navigator

2. **Who didn't speak?**

   o Did you notice?

   o Did you create space for them?

3. **What decisions were made?**

   o Were they clear to everyone?

   o Was the decision-making process transparent?

4. **What actions resulted?**

   o Clear owners?

   o Clear deadlines?

   o Clear success criteria?

Facilitation Scenarios:

Scenario 1: Strategy session with 8 senior leaders who all have strong opinions

- How do you structure it?

- How do you ensure all voices are heard?

- How do you move to decisions?

Scenario 2: Two team members are in heated disagreement, positions hardening

- How do you intervene?

- What's your facilitation approach?

- How do you move toward resolution?

Scenario 3: Meeting to solve a complex problem, but people keep rehashing old debates

- How do you redirect?

- What structure would help?

- How do you move forward?

Advanced Facilitation:

Can you facilitate:

- ☐ A strategy workshop (divergent thinking)

- ☐ A decision-making session (convergent thinking)

- ☐ A conflict resolution (emotional intelligence)

- ☐ A retrospective (psychological safety)

- ☐ A planning session (structured thinking)

Action Items:

- ☐ Design your next important meeting using a facilitation framework

- ☐ Practice asking 3 questions before making 1 statement

- ☐ Get feedback: "How effective am I at drawing out diverse perspectives?"

Show Proof of Value

The Principle: Pivotal Leaders don't just create value, they make value visible, measurable, and undeniable to skeptics.

Rate yourself (1-5):

☐ I can quantify the business impact of my team's work

☐ I tell compelling stories about value that resonate with different audiences

☐ I proactively share wins with stakeholders before they ask

☐ I connect technical achievements to business outcomes

☐ I build cases for investment with clear value propositions

☐ I celebrate and communicate value creation consistently

Value Communication Assessment:

For your team's most recent significant delivery, can you articulate:

- **Financial**
  Revenue impact: $_______ (revenue generated, protected, or enabled)
  Cost impact: $_______ (costs reduced, avoided, or optimized)
  Time impact: _______ (time saved, speed increased)

- **Governance**

  Risk impact: What risk was mitigated?

- **Strategy**

  Capability impact: What capability was enabled?

  Strategic impact: What business outcome was enabled?

If you can't quantify or tell a compelling story for at least 2 of these, you're not showing proof of value effectively. Are you relying on Proof of Concepts? Don't. Develop a Proof of Value Plan.

Value Storytelling:

Take your team's biggest win from the past quarter.

Now tell three versions:

1. **For the CEO** (60 seconds, business outcome focus)

2. **For a peer technical leader** (2 minutes, technical achievement + business context)

3. **For your team** (5 minutes, their contributions + impact + what's next)

Which version is easiest for you? Which is hardest? Why?

Proof of Value Scenarios:

Scenario 1: Your team shipped a major technical improvement. Usage is up 15%, but executives don't understand why it matters.

- How do you translate technical win to business value?

- What metrics do you highlight?

- What story do you tell?

Scenario 2: You need budget for platform work that won't show immediate user impact.

- How do you build the value case?

- What proof points do you use?

- How do you address the "why now?" question?

Scenario 3: A project delivered on time but missed the business outcome.

- How do you communicate this?

- What value can you still show?

- What do you learn and share?

Value Visibility Strategy:

- [ ] Do you have a regular cadence for sharing value?

- [ ] Do you use multiple formats (presentations, dashboards, stories)?

- [ ] Do you tailor value messages to different stakeholders?

- [ ] Do you celebrate value creation with your team?

- [ ] Do you build a "value portfolio" you can reference when needed?

Action Items:

- Quantify the business value of your team's last 3 major deliveries

- Create a "value dashboard" you update monthly and share with stakeholders

- Practice your 60-second elevator pitch of your team's value; test it with someone outside tech

Outcome-Centric (Measure & Manage To)

The Principle: Outcomes are the results that have impact. Outputs are what you produce. Pivotal Leaders manage to outcomes, not just outputs.

Rate yourself (1-5):

☐ I distinguish clearly between outputs (what we build) and outcomes (what changes)

☐ My team's goals are written as outcome statements, not task lists

☐ I measure leading indicators that predict outcomes, not just lagging results

☐ I adjust course based on outcome data, even if output metrics look good

☐ I hold myself and my team accountable to outcomes, not just effort

☐ I say "no" to outputs that don't drive outcomes

Output vs. Outcome Test:

For each statement, mark whether it's an Output (O) or Outcome (OC):

- "Ship the mobile app redesign"
- "Increase mobile conversion rate by 15%"
- "Deploy the new API"
- "Reduce API latency by 200ms"
- "Enable third-party developers to integrate"
- "Launch 10 new features"
- "Improve customer satisfaction score from 7.2 to 8.0"
- "Complete the cloud migration"
- "Reduce infrastructure costs by $2M annually"

Answers: O, OC, O, O, OC, O, OC, O, OC

If you scored less than 8/9 correct, you may be conflating outputs with outcomes.

Outcome Accountability:

Review your team's current goals or OKRs:

1. **What percentage are outcome-focused vs. output-focused?**

   - Outcome-focused: _______%

   - Output-focused: _______%

2. **For each output-focused goal, ask:**

   - What outcome does this output enable?

   - How will we know if it worked?

   - What if we delivered it and nothing changed?

Leading vs. Lagging Indicators:

For your team's top 3 outcomes, identify:

Outcome	Lagging Indicator	Leading Indicator	Frequency
Example: Increase positive impact of features deployed	Support Tickets	Direct customer surveys	Bi-weekly to Monthly

1.

2.

3.

Action Items:

- Rewrite one output-focused goal as an outcome-focused goal

- Identify leading indicators for your top 3 outcomes

- Review dashboards—what % is outcomes vs. outputs?

Be a Remedy (to the Maladies)

The Principle: Organizations have dysfunction patterns. Pivotal Leaders diagnose maladies and become the remedy through systematic root cause solutions.

Organizational Malady Diagnostic:

Guidance for dealing with many common syndromes are outlined in "The Pivotal Tech Leader's Guide to the Untraveled Road"

- Shiny Object Syndrome (SOS):
- Cobbler's Children Syndrome
- Not Invented Here (NIH) Syndrome
- Shiny Object Syndrome (SOS):
- Student Syndrome & Parkinson's Law
- The "Long No" aka Analysis Paralysis

In addition, you can check for the presence of any of these in your environment:

☐ **Hero culture**: Success depends on unsustainable individual effort

☐ **Blame culture**: Energy on fault-finding, not problem-solving

☐ **Silo mentality**: Local optimization, global suffering

☐ **Meeting culture**: Calendar full, work happens after hours

☐ **Risk aversion**: Fear prevents experimentation

☐ **Technical debt**: Short-term wins, long-term problems

☐ **Talent drain**: Best people leave

☐ **Strategic whiplash**: Direction changes quarterly

The Pivotal Leader's Companion on the Untraveled Road

Rate yourself (1-5):

☐ I diagnose organizational dysfunction, not just treat symptoms

☐ I address root causes rather than applying band-aids

☐ I create systems that prevent recurring problems

☐ I model behaviors I want to see

☐ I address cultural issues directly

☐ I know when I'm part of the problem

Action Items:

- Identify one organizational malady you can influence

- Diagnose the root cause, not just the symptom

- Take one action this week to model the remedy

Develop Allies & Partners

The Principle: A Pivotal Leader's work requires coalitions. Your leadership network is critical infrastructure.

Rate yourself (1-5):

☐ I cultivate relationships across boundaries

☐ I invest before I need them

☐ I create value without expecting immediate return

☐ I have strong cross-functional relationships

☐ I maintain industry connections

☐ I create opportunities for others

The Pivotal Leader's Companion on the Untraveled Road

Leadership Network Architecture:

Segment	Strong	Weak	Gaps
Internal - Peers			
Internal - Senior			
Internal - Other Functions			
External - Industry			
External - Advisors			

Action Items:

- Reach out to 3 allies, no agenda
- Make one introduction benefiting others
- Have coffee/lunch/a break with peer in different function

Grow Your Team

The Principle: The ultimate leadership measure is whether your team becomes more capable over time.

Rate yourself (1-5):

☐ I know each member's career aspirations

☐ I create tailored development opportunities

☐ I give growth-accelerating feedback

☐ I delegate to develop, not just offload

☐ I celebrate learning, not just results

☐ I create advancement pathways

Team Growth Assessment:

Name	Aspiration	Recent Growth	Next Opportunity	Last Discussion

Action Items:

- Career conversation with each direct report this quarter
- Create one stretch assignment for growth
- Ask team: "What do you want to learn?"

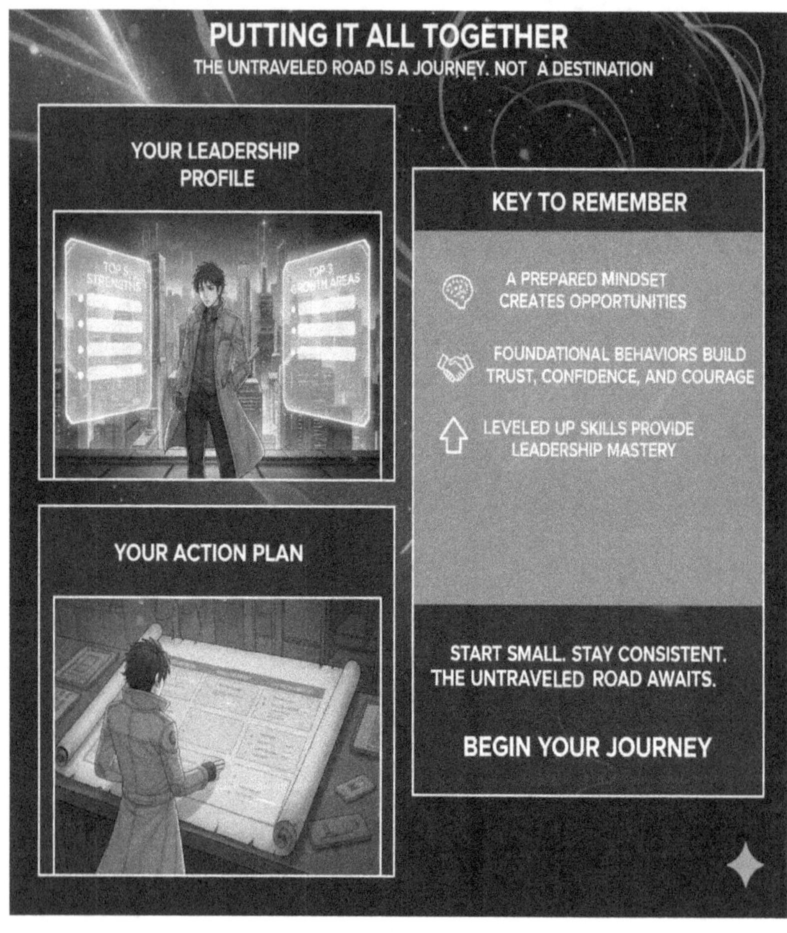

PUTTING IT ALL TOGETHER
THE UNTRAVELED ROAD IS A JOURNEY. NOT A DESTINATION
YOUR LEADERSHIP PROFILE
TOP 3 STRENGTHS
TOP 3 GROWTH AREAS
KEY TO REMEMBER
A PREPARED MINDSET CREATES OPPORTUNITIES
FOUNDATIONAL BEHAVIORS BUILD TRUST, CONFIDENCE, AND COURAGE
LEVELED UP SKILLS PROVIDE LEADERSHIP MASTERY
YOUR ACTION PLAN
START SMALL. STAY CONSISTENT. THE UNTRAVELED ROAD AWAITS.
BEGIN YOUR JOURNEY

Your Leadership Profile

Top 5 Strengths:

1. ____

____

2. ____

3. ____

4. ____

Top 3 Growth Areas:

1. ____

2. ____

3. ____

____

Your Action Plan

Next Steps

1. Choose 1-2 action per Key Aspect to improve on over the next 30/90/180 days

2. Schedule accountability check-in

3. Begin your journey

30-Day Actions:

- Mindset: _______________________________________

- Behavior: _______________________________________

- Skill: _______________________________________

90-Day Goals:

- Mindset: _______________________________________

- Behavior: _______________________________________

- Skill: _______________________________________

6-Month Transformation:

- Where I am: _______________________________________

- Where I want to be: _______________________________________

- How I'll know: _______________________________________

Accountability

Team

Check-in Frequency: _______________________

Commitments: _______________________________________

Mentors

1. ____

2. ____

Check-in Frequency: _______________

Commitments: _______________________________________

Allies

1. ____

2. ____

Check-in Frequency: _______________

Commitments: _______________________________________

Partners:

1. ____

2. ____

Check-in Frequency: _______________

Commitments: _______________________________________

Conclusion

The Untraveled Road isn't a destination; it's a continuous path of self-discovery.

This assessment reveals where you are today. Pivotal Leadership isn't about perfection, It's about intentional development, honest self-awareness, and consistent growth.

Your scores matter less than your commitment. The hardest questions are probably the most important. Your lowest areas might be your greatest opportunities.

Remember:

- A **Prepared Mindset** creates opportunities

- **Foundational Behaviors** build trust, confidence and courage

- **Leveled Up Skills** provide leadership mastery

Start small. Stay consistent. Share your journey.

The **Untraveled Road** awaits.

What will you work on first?

You are not alone on this journey.

Hit me up on any social media watering hole where we find our virtual selves.

Appendix 1: Scoring Guide

Ranking Interpretation:

5: Strength. Leverage it but watch for complacency.

4: Strong. Push for consistency.

3: Solid. Identify barriers to consistency.

2: Growth opportunity. Others likely notice.

1: Priority. Affecting your impact.